Auburn Cord Duesenberg Museum
presents

AUBURN AUTOMOBILES
1900 THROUGH 1936
PHOTO ARCHIVE

Jon M. Bill

Iconografix
Photo Archive Series

Iconografix
PO Box 446
Hudson, Wisconsin 54016 USA

The information in this book is true and complete to the best of our knowledge. All recommendations are made without any guarantee on the part of the author or Publisher, who also disclaim any liability incurred in connection with the use of this data or specific details.

We acknowledge that certain words, such as model names and designations, mentioned herein are the property of the trademark holder. We use them for purposes of identification only. This is not an official publication.

Iconografix books are offered at a discount when sold in quantity for promotional use. Businesses or organizations seeking details should write to the Marketing Department, Iconografix, at the above address.

Library of Congress Card Number: 2002112574

ISBN 1-58388-093-3

03 04 05 06 07 08 09 5 4 3 2 1

Printed in China

Cover and book design by Shawn Glidden

Copyediting by Suzie Helberg

COVER PHOTO: 1935 Auburn 851 Speedster. With its dazzling body designed by Gordon Buehrig, the 851 Speedster stands as testimony to Auburn's classic, timeless style. Speedsters had the power to fit their bold image. Each one of the supercharged thoroughbreds left the factory with a dashboard plaque certifying that the car had been driven at a speed exceeding 100 mph.

BOOK PROPOSALS

Iconografix is a publishing company specializing in books for transportation enthusiasts. We publish in a number of different areas, including Automobiles, Auto Racing, Buses, Construction Equipment, Emergency Equipment, Farming Equipment, Railroads & Trucks. The Iconografix imprint is constantly growing and expanding into new subject areas.

Authors, editors, and knowledgeable enthusiasts in the field of transportation history are invited to contact the Editorial Department at Iconografix, Inc., PO Box 446, Hudson, WI 54016.

DEDICATION

This book is dedicated to the memory of my parents, Melvin and Ruth Bill, who lived and loved in the era of the Auburn Automobile Company.

ACKNOWLEDGMENTS

The author would like to thank the following staff members of the Auburn Cord Duesenberg Museum for their support and assistance: Robert Sbarge, Museum President; Laura Brinkman, Director; Matt Short, Curator; Marcia Souers, Museum Store Manager; Gregg Buttermore, Publicist; and Norb Adams, Volunteer.

ABOUT THE AUTHOR

Jon Bill is the Director of Education and Archives at the Auburn Cord Duesenberg Museum in Auburn, Indiana. He lives in his hometown of Fort Wayne, a city 20 miles to the south. A graduate of Purdue University, majoring in Industrial Education, Jon taught Technology Education for 34 years before joining the museum staff in June of 2001.

INTRODUCTION

The quaint and charming town of Auburn, tucked away in the northeast corner of the State of Indiana, was the birthplace of no fewer than 10 different automotive nameplates during the first part of the twentieth century. Home to approximately 5,000 hard-working souls, many of them recent European immigrants, Auburn in the early 1900s was known as "The City of Honest Endeavor." However, by 1916 all but one of the automobile companies was defunct. The lone survivor was the Auburn Automobile Company, an outgrowth of the Eckhart Carriage Company. The Auburn Automobile Company, in spite of its small size, turned out some of the world's most impressive automobiles.

Although not incorporated until August of 1903, the Auburn Automobile Company was established in 1900 to investigate the feasibility of manufacturing horseless carriages. The company was capitalized with only $2,500. Charles Eckhart, also founder of the Eckhart Carriage Company, was president of the company, while his son, Morris, was named vice president and general manager. For the first few years, Morris, along with his brother Frank, studied the vehicles of various manufacturers. Eventually, after experimenting with several prototypes, the company was ready to offer its product to the public. The *Cycle and Automobile Trade Journal* reported in its April 1, 1903, issue that an Auburn gasoline car was available for a price of $1,400 including side lamps and a tool kit. Sales reportedly totaled approximately 100 during the first year of production, however, the exact number is impossible to ascertain. The first Auburns were powered by a single-cylinder engine.

Encouraged by the modest success of the 1903 and subsequent 1904 models, the Eckharts expanded production facilities. No longer would Auburns be built in the carriage factory; instead a new two-story structure was erected behind the carriage complex for the sole purpose of automobile production. Auburn was now poised to make its mark in the fledgling automobile industry. Sales of Auburns increased every year through 1910 when 1,365 cars were built. By 1916 production had nearly doubled to 2,686.

Throughout the company's short but glorious history, Auburn automobiles had a well-earned reputation as a car capable of extraordinary performance and endurance. In 1910 Melvin Leasure, factory superintendent, won the 40-horsepower-class stock car race in an Auburn at the Oklahoma State Fairgrounds, covering five miles in seven minutes and five seconds, for an average speed of 42 mph. Speed and endurance would be recurring themes in days to come.

One and two cylinder Auburns through 1911 were powered by an engine of the company's own design and manufacture. During the 1909 model year, four-cylinder engines built by Rutenber of Logansport, Indiana were introduced in some models. Rutenber engines were used in both four and six cylinder form until 1916 when they were replaced by Teetor engines (1915-1917) and Continental engines (1915-1927). Engines manufactured by Lycoming debuted in the Auburn line in 1925 and by 1928 were used exclusively by Auburn until the end in 1936, an exception being a few Cummins Diesel powered cars built for testing and demonstration purposes.

Following the death of the Eckhart family patriarch Charles, in late September 1915, his sons decided to put the company up for sale. However, they found no takers until 1919 when a group of investors from Chicago bought the Auburn

Automobile Company. This group, which included William K. Wrigley, Jr., the chewing gum magnate, may have been wise in the world of finance, but they were neophytes in the automobile business. The Auburn "Beauty Six," designed under the Eckhart's management and introduced in 1919, was a very handsome car that set a new sales record for the company at 6,062 units. However, due to the post-war recession, sales fell for the next three years, reaching a low of 2,408 in 1922. The Chicago group was justly concerned about shrinking profits. From well over $1,000,000 in 1919, profits fell every year through 1923 when they reached a meager low of $8,342. In 1924 the company recorded a deficit of nearly $70,000. To make matters worse, in a field behind the factory stood 700 stagnant, unsold Auburns. The investors realized that something had to be done.

Enter E. L. Cord, who was to make an immediate and lasting impact on the transportation industry in general and on the Auburn Automobile Company in particular.

Errett Lobban Cord, named after the clergyman who married his parents and his mother's maiden name, was born in Warrensburg, Missouri on July 20, 1894. Cord did not particularly care for his given name and preferred to be called "E. L." Cord was a very bright youngster. His mother was a teacher and he inherited her love of learning. When asked what he wanted for his fourth birthday, Cord replied, "I wish I could read." Cord's family moved from time to time and they eventually settled in Los Angeles when Cord was in his teens. He was a capable student but perhaps because of his brilliance, lost interest in school. At the age of 16, he ended his formal education when high school classes concluded for the summer of 1909. In 1911 his father died and Cord took his first job as a used car salesman. He would later say, "Just like any other kid, I was crazy about cars." By 1914 Cord was working in a service station by day and using the owner's shop at night to build racecars. Cord would buy a well-used Ford Model T at a low price, replace the original body with one that was lightweight and sleek, modify the engine for high-performance and take his creation racing. When the car was a proven winner, he would sell it for a tidy profit as winning cars commanded top dollar. Cord perceived that cars with eye-appealing style and powerful engines would essentially sell themselves. This concept would serve him well in the next two decades.

When interviewed in later life, Cord remarked that a conquered challenge was "no longer fun. You want to tackle something else. All you can do by staying where you are is make more money, and that isn't fun." Throughout his working career, these were words by which Cord lived. So he was off to other ventures, some successful and some not. Cord was to accumulate and lose several small fortunes before his arrival in Chicago in 1920 as a salesman with the Moon Motor Car Company. With the energetic Cord on board, sales skyrocketed with the Chicago region accounting for 60 percent of all Moons sold. By 1924, Cord's sales success had caught the attention of the Chicago investors that owned the moribund Auburn Automobile Company. One of the investors, Ralph Bard, arranged to purchase a Moon from Cord, using the opportunity to woo him with an offer from Auburn. A generous contract was tendered but Cord turned it down. Instead, he asked for less salary or 20 percent of the profits - whichever was greater, the option to buy all common stock, plus total decision-making control. If successful, he could then be in position to buy out the Chicago investors and the company would be his. At first the owners balked at the deal, but then realized that currently there were no profits. If this brash young man could do what he said, at least they would recover their investment. Cord moved to Auburn during the summer of 1924 and things began to happen in short order. The 700 Auburns languishing in the field behind the factory were treated to new bright two-tone paint jobs. Prices were slashed, a persuasive national advertising campaign was launched, and dealers, encouraged by Cord's enthusiasm, sold the cars to an eager public. Cord would later say that the ad campaign was a "goddam wow." The capital generated by the sale allowed Cord to pursue the design of his dream car.

In January of 1925 Auburn introduced the lowest priced eight-cylinder car in the country. Using an engine manufactured by Lycoming, the straight-eight Auburn hit the showrooms at a bargain price of $1,895 for a touring car. At last the buyer of a medium-priced car could enjoy the smoothness, performance, and luxury of eight-cylinder power. But this car was merely a transition model until the new Cord-influenced 6-66 and 8-88 models were introduced midyear. Sales jumped and in November of 1925 Cord paid off the Chicago investors and assumed ownership of the company. At the age of 31 Cord embarked upon building his transportation empire.

Within a few short years he would own Duesenberg Automobile and Motors Co., Lycoming Manufacturing, Spencer Heater Co., Checker Cab, Columbia Axle, Stinson Aircraft, Century Air Lines, two auto body manufacturing plants and an additional auto assembly plant among other holdings. Eventually these all fell under the umbrella of the Cord Corporation, E. L. Cord's holding company.

Under Cord's direction, Auburn sales continued to grow. In 1925 production totaled 5,600 cars. Three new factory buildings increased production capacity to 150 cars per day. During 1926 Auburn sales set a new company record at 8,500 units sold. Net profits were $800,000. Cord realized that Auburn could never compete on an even field with the likes of Ford, General Motors, or Chrysler so he looked for a niche in the auto industry where he could specialize and offer something unique. He was quoted as saying, "If you can't be the biggest, it pays to be different." A prophetic statement indeed as originality and innovation would play major roles in Auburn's success.

The "Roaring Twenties" continued to roar for Auburn in terms of popularity. Production, boosted in part by export sales, increased to nearly 14,000 units for both 1927 and 1928. In March of 1927, noted racecar driver Wade Morton set a new speed and endurance record in an Auburn 8-88 at the Atlantic City Motor Speedway on its 1.5-mile high-banked board track. Morton's roadster covered 10,000 miles with an average speed of over 72 mph for the last 1,000 miles. This feat brought people into Auburn showrooms throughout the nation and firmly established a long-lasting reputation for high-performance. In 1928, Auburn debuted its most famous production body style, the Speedster. With its boattail stern, air flowed smoothly around the body creating less drag. Using a straight-eight engine, newly designed for 1928, another speed record was set by Wade Morton at Daytona Beach at over 108 mph for the flying mile. In 1929 Auburn sales soared to over 20,000.

During 1930 sales fell to 11,357 as Auburn was in the final year of a three-year body style as well as feeling the effects of the stock market crash. However, in 1931 an all time sales record was set when a beautifully redesigned Auburn was introduced. Former Duesenberg body stylist, Alan H. Leamy, created a stunningly gorgeous design that won wide acceptance by the car buying public. Auburn cut its model offerings to one, the 8-98, which stood for eight cylinders and 98 horsepower. At the end of the model year over 32,000 Auburns were sold, good enough for thirteenth place among American auto manufacturers. Profits stood at $4.1 million.

About this time, E. L. Cord left the day-to-day operation of Auburn to his executives in order to pursue other interests. It was to be Auburn's last good year.

The year 1932 saw Auburn continue its theme of innovation and value. A new V-12 Lycoming engine rated at 160 horsepower debuted at the astoundingly low price of $1,425 for a coupe. By comparison, the next lowest priced V-12 coupe was Cadillac at $3,395. A new two-speed rear axle was also introduced. Termed "Dual-Ratio," it allowed the driver to control the rear axle ratio by simply turning a knob on the dashboard. With a setting for power (low) and a setting for speed (high), Dual-Ratio permitted both performance and economy. It was a great sales tool during the Depression. However, despite these innovations and a midyear price cut, Auburn sales tumbled to 11,000, barely one-third of the previous year. Profits turned into a loss of $1.1 million. The year 1933 was even more dismal. In the final year of the 1931 body style, the 1933s now looked stale, with the exception of the face-lifted, top-line Salon models which few people could afford. Fewer than 5,000 were sold and Auburn continued to lose money.

In 1934, Auburn introduced what was to be its final body style. The bodies were now of all-steel construction with tasteful styling by Leamy. Auburn offered a six-cylinder engine for the first time since 1930. Sales showed a modest improvement to over 7,700 units but it wasn't enough to stem the tide of red ink. Under these grim conditions, management brought in another young and talented designer from Duesenberg, Gordon Buehrig. His task was to facelift the Auburn for 1935. With a miniscule budget, Buehrig managed to create some of the most beautiful Auburns ever built. High-performance continued to be an Auburn byword in 1935 with the introduction of a supercharged eight-cylinder engine with a rating of 150 horsepower. By comparison, the standard Auburn eight-cylinder engine produced 115 and a Ford V-8 produced 85. Racecar driver Ab Jenkins ran a supercharged Auburn Speedster at the Bonneville Salt Flats, setting 70 new American and international records. However, sales fell once again to slightly over 6,300 units for the 1935 model year. The 1936 Auburns were little more than rebadged, leftover 1935 models. Sales in 1936 plummeted to 1,250 units, the lowest since 1914. The end for the struggling independent was at hand, a victim of the Depression and changing public tastes. Auburn production ended with a grand total of over 177,000 cars manufactured. In 1937, E. L. Cord sold the Cord Corporation empire for $2.6 million and moved to his West Coast home where he invested in real estate, mines, broadcasting stations, and oil wells.

Thirty-six short years saw Auburn rise and fall as one of America's most innovative automobile manufacturers. During that time Auburn always offered an in-vogue, quality product at a reasonable price. The Auburns that remain today stand as testimony to the company's fine engineering and styling which continue to influence the design of modern automobiles.

The photographs appearing in *Auburn Automobiles 1900 through 1936 Photo Archive* are from the collection of the Auburn Cord Duesenberg Museum. Established in 1974, the museum is located at 1600 South Wayne Street in Auburn, Indiana. It occupies the original Auburn Automobile Company factory showroom and administration building, which has been elegantly restored to its former art deco splendor. The museum honors three extraordinary marques from America's automotive golden age and exhibits 100 automobiles in themed galleries. The museum is open daily, year-round. Web site: www.acdmuseum.org

Jon M. Bill, Director of Education and Archives
Auburn Cord Duesenberg Museum

1902 Auburn Runabout Prototype. After formation of the Auburn Automobile Company in 1900, brothers Frank and Morris Eckhart experimented by building several prototypes. The first successful runabout was completed in 1902. It carried two passengers and used a water-cooled gasoline engine. While many early autos used tiller steering, production Auburns used steering wheels from the beginning.

1902 Auburn Runabout Prototype. This rear view shows the chain drive mechanism used on early Auburns as well as the three-quarter elliptical springs on both the front and rear axles. The engine was positioned under the driver's seat. This prototype was the basis for the 1903 production model, which used a four-stroke, 95-cubic-inch, single-cylinder engine, two-speed planetary transmission, and a 78-inch wheelbase.

1904 Auburn Model A, Rear-Entrance Tonneau. Back seat passengers entered through a small door on the rear of the seat. Springs on this car are semi-elliptical in front and full elliptical in the rear. List price was $1,250. This car is the oldest Auburn known to exist and is on exhibit at the Auburn Cord Duesenberg Museum.

1906 Auburn Model C Touring. By 1906 Auburns were powered by a two-cylinder opposed engine developing approximately 20 horsepower at 800 rpm. The Model C weighed 1,750 pounds and had a 94-inch wheelbase. Top speed was reported to be 40 mph. Entrance to the back seat was made much easier through doors on the side. This model was the only body style Auburn offered in 1906. In 1907 the factory catalog listed two touring cars with differing levels of equipment and a runabout. Auburn's slogan was "The Most for the Money."

1908 Auburn Model H Touring. The Model H was an upgrade from the standard Model G. It came equipped with a black top, wider tires, gas lamps, tail light, and storage battery. A runabout model was also available. The box on the running board was used to carry tools. The five-passenger auto pictured here was used to promote the 1923 Auburn line. In advertisements homage was frequently paid to Auburn's past.

1908 Auburn Model G Touring. These gentlemen had just returned to Auburn from Chicago in November of 1907, where they finished fifth among 35 cars in a reliability run sponsored by the Chicago Motor Club. This car was equipped with optional gas headlamps, which supplemented the kerosene side lamps. In 1908 the wheelbase grew to 100 inches, weight was up to 2,000 pounds, and horsepower was rated at 24. List price for the Model G was $1,350.

1910 Auburn Model X Touring. The 1910 Auburn line-up expanded to six models, four of which were powered by a Rutenber four-cylinder engine introduced in 1909. Body styles available were touring, baby tonneau, and roadster. The four-cylinder models were equipped with a three-speed sliding gear transmission and bevel gear differential. This photo appeared in the 1910 catalog with the caption "House Party at Coldwater Lake, Michigan."

1910 Auburn Model S Roadster (Stripped for racing). In April of 1910 Auburn's first racing victory was recorded at the Oklahoma State Fairgrounds' half-mile dirt track. With factory superintendent Melvin Leasure at the wheel, the 40 horsepower racer covered five miles in seven minutes and five seconds for a winning average speed of over 42 mph. The Oklahoma City newspaper reported that "Persistent and constant work made Auburn an easy winner, defeating cars up to twice its price."

1910 Auburn Four-Cylinder Engine. This is the engine that powered Auburn to its first racing victory. Built by Rutenber of Logansport, Indiana, the engine had four iron cylinders cast separately. The crankcase, formed of two horizontal sections, was cast of aluminum alloy. The forged steel crankshaft was supported by five bearings. The bore and stroke were 4 1/2 x 5 inches, respectively, for a displacement of 318 cubic inches. Peak horsepower was rated at 40. The exposed tappets on the left side of the engine provided easy valve adjustment. The Remy magneto was positioned on the right side. Auburn used this engine in various models through 1915.

Motor Four Cylinder

1911 Auburn Model L Fore (sic) – Door. Auburn's first model with a front door. Only a front door was provided on the left as control levers located on the outside of the right-hand drive required a fixed panel on that side. The front seat passengers' legs were now protected from the wind. The Model L used a four-cylinder engine producing 25 to 30 horsepower and a 112-inch wheelbase. List price was $1,400. The year 1911 would be the last for two-cylinder Auburns. The Double Fabric Tire Company factory was located in Auburn, Indiana and used this car for promotional purposes.

1911-1912 Auburn Model N Touring. Two factory workers pause to have their picture taken while preparing to load a new Auburn into a railroad boxcar on a cold winter day. The Model N rode on a 120-inch wheelbase and in 1911 was the top of Auburn's line. Its 40 horsepower engine provided plenty of go. Coming to a stop was a different matter, with mechanical brakes operating on the rear wheels only.

1912 Auburns. Three Auburns from the J. J. Meyer Agency based in East Orange, New Jersey are on display at the New York Auto Show. New for 1912 was a six-cylinder engine that was rated at 50 horsepower. It was used in only one model, a touring car appropriately named the Six-50 (first and third cars from left in photo). Listing at $3,000, it had a seven-passenger capacity and was the first Auburn to be equipped with electric lights.

1913 Auburn Model 40 Town Car. Big news for Auburn in 1913 was the debut of three models of enclosed cars; a coupe, limousine, and town car. The complete Auburn line included 14 models ranging in price from $1,150 to $3,200. The town car had a four-cylinder, 40 horsepower engine with a bore of 4 1/2 inches and a stroke of 5 inches. Gasoline was fed by gravity from an 18-gallon tank positioned under the front seat. The town car had two doors in staggered locations. The right door opened into the rear seat area and the left door opened into the front seat area.

1913/1914 Auburn Touring. Two Harley-Davidson motorcycle riders pose for the camera with the occupants of an Auburn Touring car. The year of this car is difficult to determine, as it has elements of 1913 (right-hand drive and a straight front fender line) and 1914 (doors with hidden hinges and stylish curve where the windshield meets the front door). The 1913 full-line catalog stated, "We prefer the Auburn cars to become known as the best cars and not as the most numerous."

1915 Auburn Touring. This car, parked in the driveway of the Auburn home of Morris Eckhart, one of the owners of the Auburn Automobile Company, may have been a prototype for the 1915 models. The hood and cowl now blend together in a smooth curve rather than the hood being butted to the cowl. By 1915, all domestic Auburns had the steering located on the left side and the gasoline tank positioned in the rear.

1922 Auburns. Auburn introduced the "Beauty Six" model in 1919 and continued the line through 1922. Body styles included a sedan, roadster, coupe, and a touring car. All models were powered by a Continental "Red Seal" six-cylinder engine of 224 cubic inches rated at 55 horsepower at 2,600 rpm. This photo was taken in the showroom of the F. A. Dutton Motor Company situated in Boston, Massachusetts.

1922 Auburn Sedan. The Auburn Beauty Six Sedan featured seating for five passengers with auxiliary seats for two more. The four-door sedan weighed in at 3,430 pounds and in January had a factory list price of $2,395, making it the most expensive model offered in 1922. Auburn sales totaled 2,408 units, the lowest number since 1918.

1922 Auburn Roadster. This photo, which appeared in the 1922 catalog, illustrates Auburn's graceful styling. Notable is the front hinged door with a vertical handle and an accent color panel at the top. The rear deck had a luggage compartment. Front and rear bumpers were an optional item. The roadster started out the year with a list price of $1,575.

A 1923 Auburn Brougham. New for 1923 was the two-door brougham. The Auburn catalog stated of this model, "The beauty of the car is in part due to the low center of gravity and the lowness of the roof; yet there has been no sacrifice of spaciousness, and the head-room is ample." Beginning in 1923, Auburn used the slogan, "Once an Owner - Always a Friend."

1923 Auburn Sport Touring. "It is pert and voguish but never so extreme as to pass the bounds of good taste," so claimed the sales catalog of the sport model. Note the use of drum-style headlamps and individual step plates rather than full-length running boards. The trunk is carried on the side instead of the rear and disc wheels were used. A cowl vent improved interior comfort. These folks were out for a round of golf at the Auburn Country Club.

1923 Auburn 6-43 Touring. Open touring cars were quite chilly in the cold winter months so a snug winter enclosure was a welcome option. The lower priced 6-43 models established Auburn's junior series. They were powered by a Continental six-cylinder engine of 196 cubic inches that developed 50 horsepower at 2,600 rpm.

1924 Auburn 6-43 English Coach. A midyear entry into the Auburn line-up was the fashionable English Coach promoted by Auburn newcomer E. L. Cord. Featuring a rounded rear and sides, with a high European-style waist, this rakish model came fully equipped with balloon tires, front bumper, and a nickel-plated radiator. Fresh air vents were built-in above the windshield under the visor.

1925 Auburn 8-63 Sedan. In January of 1925 E. L. Cord introduced Auburns powered by an eight-cylinder Lycoming engine. Most cars in the Auburn's price range used six-cylinder engines. The catalog encouraged prospective buyers to "step on the gas - tramp on it hard if you dare - its smoothness of performance makes you want to tell your friends." The list price was $2,550 for the sedan.

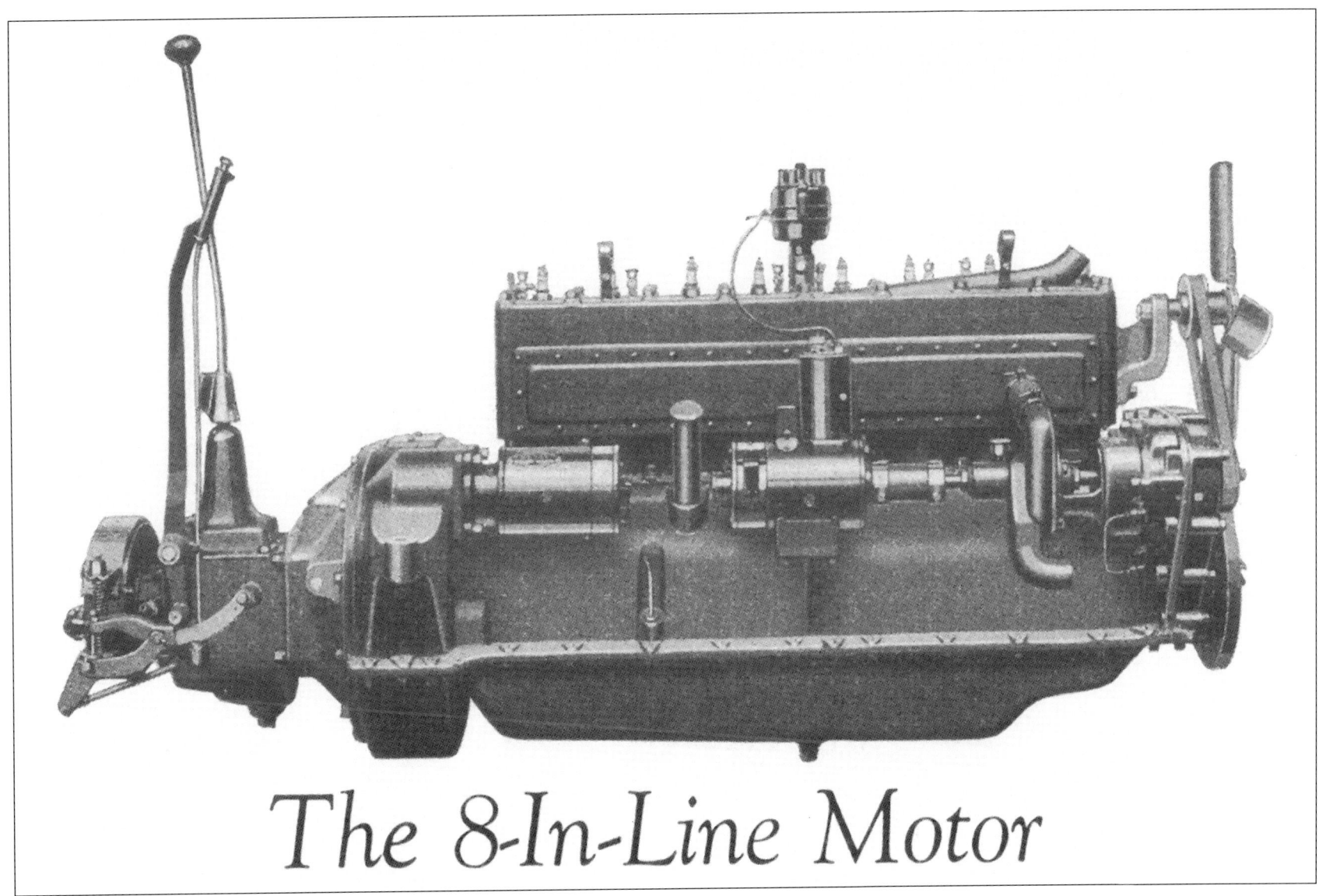

The 8-In-Line Motor

1925 Auburn Eight-Cylinder Engine. This photo taken from original factory literature illustrates the engine that established Auburn's reputation for high-performance. It was built by Lycoming Manufacturing Company of Williamsport, Pennsylvania. With a bore and stroke of 3 1/8 x 4 1/4, it displaced 261 cubic inches. Horsepower was rated at 58 at 2,850 rpm. A pump circulated motor oil with pressure controlled by the opening and closing of the throttle, not by the speed of the motor.

1926 Auburn 8-88 Sedan. The brilliant industrialist E. L. Cord (left) poses with vice-president and business manager Roy Faulkner by a new Auburn Eight-cylinder Sedan at the Auburn Automobile Company plant. Faulkner, who started with the company in 1922, would later rise to the position of President of Auburn in 1931 when E. L. assumed the position of Chairman of the Board. The 8-88, introduced in mid-1925, was Cord's brainchild and represents the first car built totally under Cord's influence.

1926 Auburn 8-88 Touring. Touted as "America's Fastest Stock Car," this Auburn reportedly covered over 400,000 miles in an around-the-world tour. This car is equipped with an unusual front bumper and has had the running boards removed. The body is covered with emblems and signatures from throughout the world.

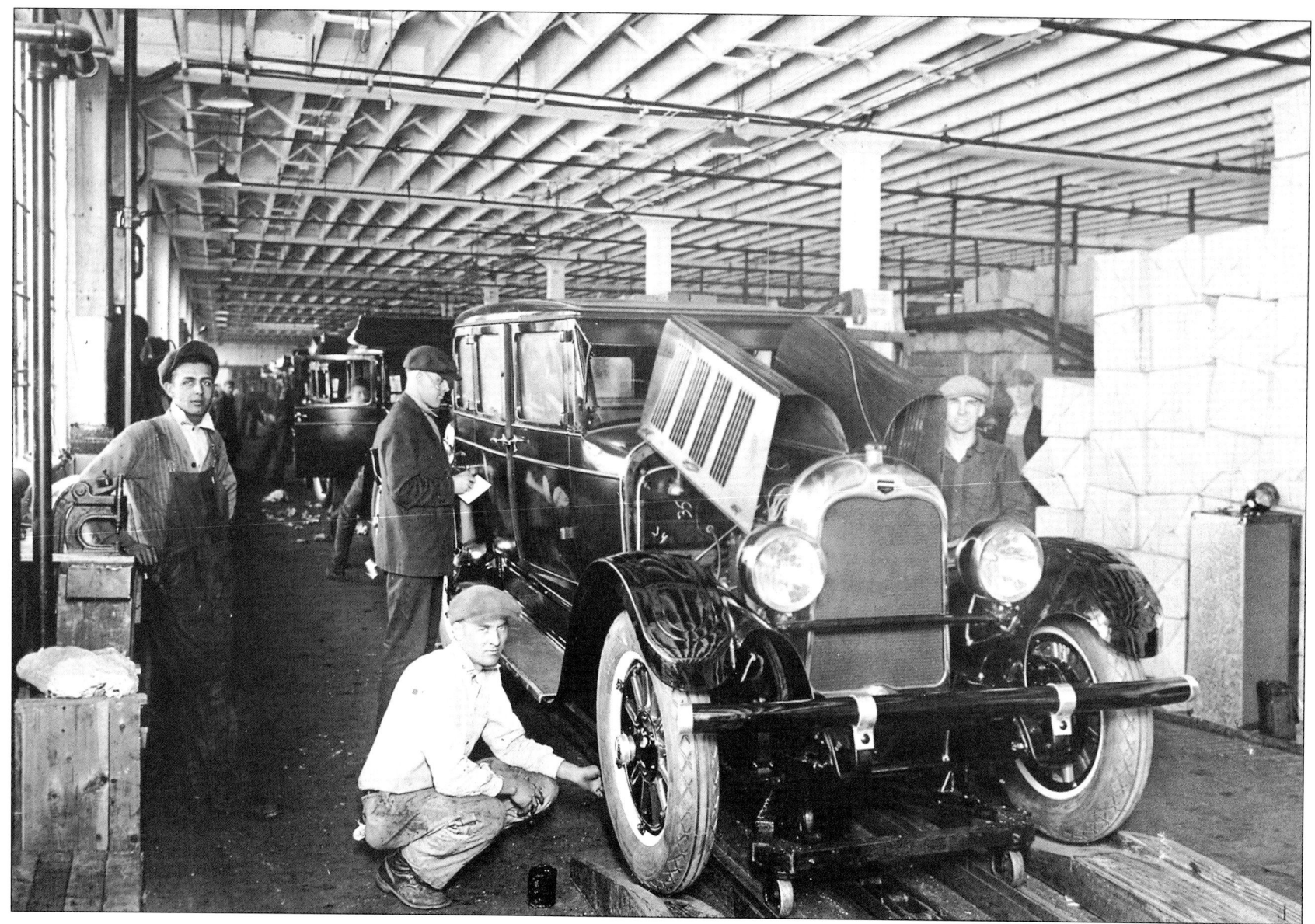

Auburn Factory - Final Assembly. Workers put on the finishing touches and inspect an 8-88 Auburn Sedan at the end of the production line. The starting pay for most factory workers was $.25 per hour. The tubular bumper was attached by "U" brackets made of spring steel, which offered protection under impact. Production capacity was 150 cars per day in this non-union plant.

1926 Auburn 4-44 Sedan Body. In 1926, for one year only, Auburn reintroduced four-cylinder engines in the 4-44 series. Customers could now choose between four, six, or eight-cylinder power. Four-cylinder Auburns rode on the same 120-inch wheelbase as the six-cylinder models. This body, with a visor unique to this model, is being finished in pyroxylin lacquer that will later be hand rubbed to a radiant shine.

1926 Auburn Quaker Oats Company Publicity Vehicle. An 8-88 chassis and portions of a touring car body were used to construct this unusual vehicle. The cannon was used to shoot breakfast cereal into the octagon-shaped container in a demonstration showing that Quaker Cereals were indeed "Shot from Guns."

1927 Auburn 8-88 Seven-Passenger Touring. Sitting on a wheelbase of 147 inches, the seven-passenger touring could carry its passengers with room to spare. It was advertised as the longest stock car in America. The car weighed in at 3,625 pounds and carried a price tag ranging from $2,245 to $2,695.

1927 Auburn 8-88 Cabriolet. Debuting midyear was this handsome cabriolet which possessed all of Cord's wishes: a colorful exterior, an attractive belt line which curved along the hood and ran the length of the body, hood louvers in forward tilting sets, and straight-eight power. The rumble seat provided room for an extra passenger or two. The small door accessed a compartment convenient for storing golf clubs.

35

1927 Auburn 8-88 Roadsters and Sedan. In March of 1927 three Auburns, two roadsters and a sedan shattered all American stock-car records for fully equipped cars. Cumulatively the trio ran 42,000 miles in less than 40,435 minutes at the 1.5-mile board track at the Atlantic City Motor Speedway under the watchful eye of the American Automobile Association. The time included all pit stops for fuel and change of drivers. Renowned racecar driver Wade Morton (wearing the beret) piloted one of the roadsters. The 8-88 series eight-cylinder engine produced 90 horsepower.

1928 Auburn 8-88 Phaeton Prototype. Auburn introduced a convertible sedan model in 1928. Not shown are the door windows, which could be raised or lowered along with the top. Owners could now enjoy the open air of a touring car in fair weather and the comfort of a sedan in foul weather. The 1928 headlamps varied from the type shown in this photo.

1928 Auburn 115 Speedster. Debuting in 1928 was Auburn's most acclaimed body style, the Speedster. The flowing, sensuous body lines speak for themselves. Lycoming created an engine worthy of such a beauty, a 115 horsepower straight-eight. Displacing 299 cubic inches, sporting a dual-type carburetor, and a 6.25 to 1 compression ratio, the Speedster's engine was well suited to further Auburn's reputation for high-performance.

1928 Auburn 115 Speedster. Auburn wasted no time in demonstrating the performance capabilities of the Speedster. Wade Morton was again in the driver's seat as a new top speed stock car record was set on the sands of Daytona Beach, Florida in February of 1928; 108.460 mph for the flying mile. The 115 horsepower developed by the engine enabled Auburn to claim that no other American production car was so powerful.

1928 Auburn 115 Cabriolet. An airmail pilot pauses with his biplane and an Auburn Cabriolet at the Auburn, Indiana airport. Aircraft were to play a major role in the future of E. L. Cord and his holdings. Knock-off hubcaps secured the optional wire wheels.

1928 Auburn 88 Speedster. Celebrities were frequently used to promote Auburn products. Pictured is the stuttering burlesque comedian Roscoe Ates in a Model 88 Speedster. Ates was a faithful Auburn owner and spokesman through the 1930s. One gentleman in the photo seems more interested in the curves depicted on the billboard than the curves on the Speedster.

1929 Auburn 115 Speedster. Musician Huston Ray finds another use for the eye-grabbing Speedster. With a miniature piano attached to the rear of the car and taking direction from the pretty lady perched atop, Mr. Ray promoted his musical production at impromptu stops. The spare tire cover reads, "Watch for Huston Ray in *The Music Healer* coming soon."

1929 Auburns. A phaeton, sport sedan, and sedan (left to right) share floor space at the Montreal Motor Show during January of 1929. The Auburn Automobile Company had a very dynamic export division with 99 dealers located in 93 countries and 3 U. S. territories. At one time Auburn exported as much as 17 percent of its production.

1929 Auburn 8-90 Sport Sedan. The distinctive feature of the Sport Sedan was the absence of rear quarter windows, which gave the rear seat passengers a sense of privacy and exclusivity. The 8-90 series used an eight-cylinder Lycoming engine of 248 cubic inches that developed 93 horsepower. The standard compression ratio was 5.15 to 1 with 6.25 to 1 offered as an option.

1929 Auburn 120 Victoria. This well-proportioned four-passenger body style was not popular with Auburn buyers and was destined to last only one year. The rear seating compartment was situated in front of the rear axle resulting in a more comfortable ride. The optional wire wheels carried 6.50 x 18 tires. The Auburn Automobile Company was approaching the pinnacle of its success in 1929 and employed over 12,000 people.

1925, 1926, 1927, and 1928-1929 Auburn Sedans. Prepared to embark on a tour of principal U. S. cities, these Auburn Sedans illustrate the company's policy of styling continuity. A minimum of annual changes was claimed to protect the owner's investment.

1929 Auburn 8-90 Phaeton. "In the Model 8-90, Auburn offers a straight-eight at prices putting it within reach of thousands of people who have previously paid more for six-cylinder cars not comparable in size nor performance," so claimed the 1929 Auburn catalog. The accessory "Pilot Ray Automatic Driving Light" was controlled by the steering mechanism.

1929 Auburns and Noble Truck. On a warm summer day, Auburn Sedans and Sport Sedans are prepared to be delivered to waiting customers, hauled behind a Noble truck. Nobles were manufactured from 1917 through 1931 in Kendallville, Indiana.

1929 Auburn 120 Phaeton. An Auburn Phaeton passes through an ancient Sequoia tree in Yosemite National Park, California. The 227-foot tree is estimated to be 2,100 years old. The roadway was cut through in 1875. The 120 series was Auburn's top line in 1929. They were equipped with four-wheel hydraulic brakes manufactured by Lockheed. The eight-cylinder engine produced 125 horsepower from 299 cubic inches. Auburn 120 frames were very sturdy, made from 3/16-inch steel, utilizing a 7-inch deep channel with 3-inch wide flanges.

1929 Auburn Cabin Speedster. E. L. Cord's fascination with aircraft was translated to a four-wheeled vehicle with the creation of this intriguing one-off show car. Griswold Motor Body Company built the aluminum over hardwood body, which was assembled to a modified Auburn frame. Note the use of motorcycle-type fenders. Woodlite headlamps were yet to be installed when this photograph was taken at the Auburn plant.

1929 Auburn Cabin Speedster. The 3,000-pound car was capable of 100 mph using the Series 120 engine of 125 horsepower. The patent drawings filed in 1929 carried the name of acclaimed racecar driver Wade Morton as the inventor. Unfortunately the car, along with 320 other cars, was destroyed in a fire at the Los Angeles Automobile Show in March of 1929.

1929 Auburn 120 Phaeton. Young ladies representing the Miss Physical Culture contest grace the lines of an Auburn 120 Phaeton. Publicity photographs such as these helped promote the automobile as well as the event.

1929 Auburn Phaeton and Cord Sedan with the Graf Zeppelin. On September 10, 1929, the German dirigible Graf Zeppelin touched down in California on its cross-country tour. The new Cord Sedan and Auburn Phaeton seem to be attracting as much attention as the airship. Both cars are equipped with the optional Pilot Ray Automatic Driving Light, a factory approved accessory.

1929 Auburn 120 Cabriolet. Cabriolets were equipped with a rumble seat, which was opened by the handle on the rear deck. Unlike the Speedster, the Cabriolet tops did not fold into the body when lowered. Open cars were upholstered in luxurious leather rather than the mohair fabric used in the closed models. The Series 120 Cabriolet listed at $1,895 and weighed 3,800 pounds.

1930 Auburn 125 Phaeton. 1930 saw Auburn's first significant reversal since E. L. Cord's arrival. Sales suffered a 71 percent decrease. This was mainly due to the economic depression and Auburn being in the final year of the 1928 through 1930 body style. The Auburn Series 125 at $1,495, was priced between the Chrysler 70 ($1,445) and Buick 50 ($1,540). The engines in the Chrysler and Buick, both six-cylinders, were rated at 93 and 98 horsepower respectively compared to Auburn's 125. The car pictured has a custom aluminum body built by the Walter M. Murphy Company.

1931 Auburn Sedan Prototype. This prototype, with its graceful fender lines, reveals the influence of the brilliant stylist Alan H. Leamy. The prototype, mounted on casters for ease of movement, has a frame constructed of wood. Note the C-clamp holding the fender in position.

Auburn Automobile Company Factory Complex. This photograph from the early 1930s shows the factory complex located in south Auburn, Indiana. The "J" shaped building in the center foreground is the Factory Showroom and Administration Building which is now the home of the popular Auburn Cord Duesenberg Museum. The building to the rear of the Administration Building presently houses the National Automotive and Truck Museum of the United States (NATMUS).

Auburn Automobile Company Administration Building. Opening in September of 1930, this exquisite building housed administrative offices, experimental, engineering, and design departments, and a magnificent showroom that displayed the products of E. L. Cord's transportation empire. Originally of a modernistic art-deco motif, the building has been painstakingly restored and now houses the world-renowned Auburn Cord Duesenberg Museum. This photograph shows seven Auburns and one Cord (left), purchased by the members of the popular Coon-Sanders Radio Orchestra in April of 1931.

Auburn Chassis Assembly. The Administration Building housed an electric chassis dynamometer used by engineers to determine the rear wheel horsepower of Auburn automobiles. To establish its rating, a lashed down Auburn chassis spins the rollers built into the floor.

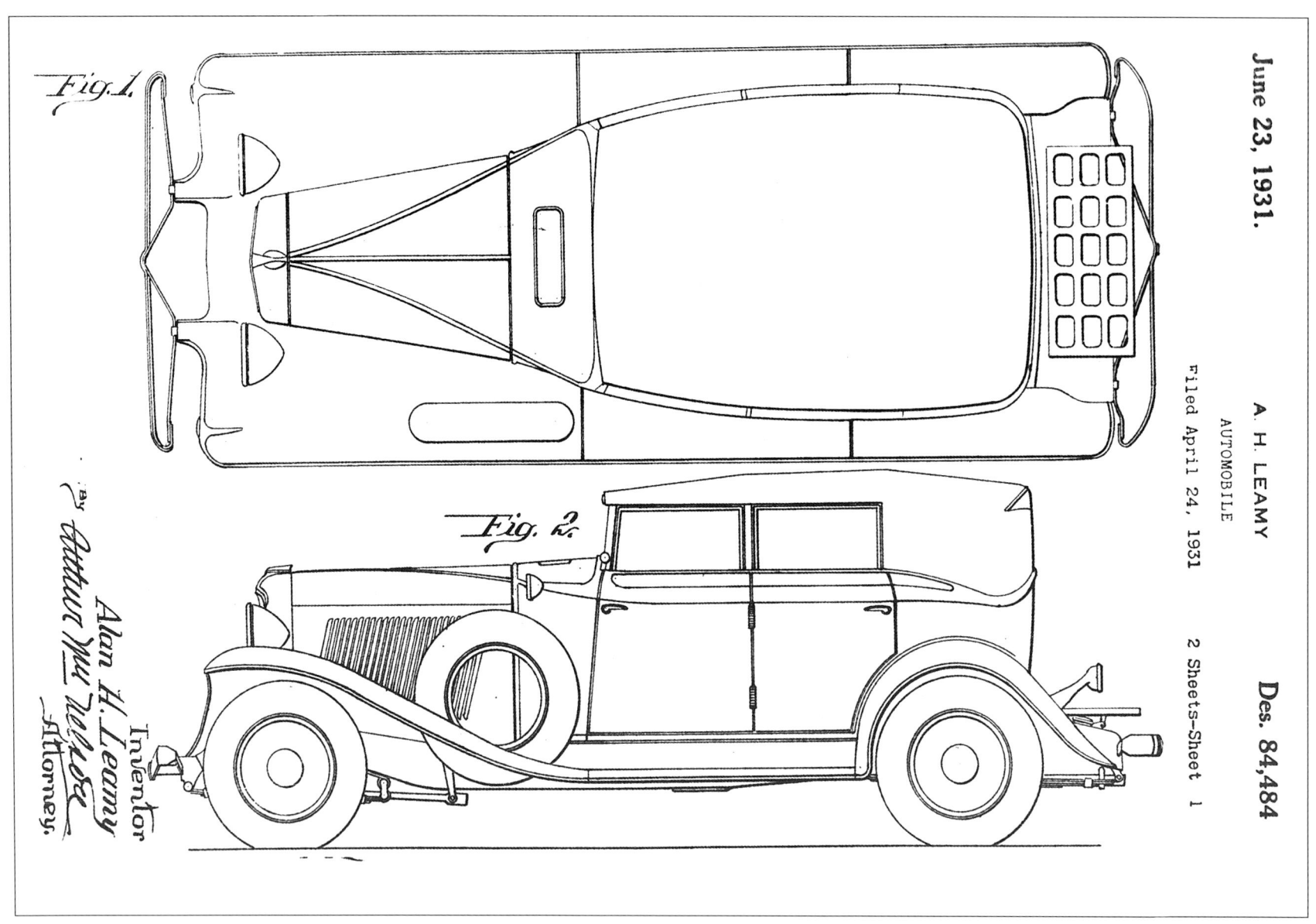

June 23, 1931.

A. H. LEAMY

AUTOMOBILE

Filed April 24, 1931

Des. 84,484

2 Sheets—Sheet 1

Fig. 1.

Fig. 2.

Alan H. Leamy
Inventor
By Arthur M. Nelson
Attorney.

1931 Auburn Phaeton Patent Drawings. The gifted stylist Alan H. Leamy, designer of the front end and fenders of the spectacular Duesenberg Model J, came to Auburn in 1928 where he was responsible for the Auburn's elegant looks from 1931 through 1934. Leamy also designed the famous Cord L-29. Stricken with polio as a youth, Leamy's life was tragically cut short at the age of 33 when he died of blood poisoning from a medically administered injection.

1931 Auburn 8-98 Speedster. Alan Leamy is pictured behind the wheel of one of his creations in front of the Auburn Automobile Company Administration Building. The Speedster body style was revived in October of 1931 after a hiatus during 1930. With the top raised over the cockpit, the Speedster measured a scant 63 inches in height.

Auburn-Cord Dealership – 1931. Edwards Motors located in Louisville, Kentucky was in for a banner year during 1931 when Auburn placed thirteenth in sales among American automobile manufacturers. A Cord L-29 Sedan (left) and an Auburn Sedan grace the street while an Auburn Phaeton is displayed in the window on the second floor. 1931 Auburn production would exceed 32,000 units.

1931 Auburn 8-98 Brougham. Auburn's dramatic front-end styling excited the car buying public. Other manufacturers took note and soon elements of Auburn designs began to appear on competitor's products. During 1931 Auburn cut their model offerings to one, available in either a standard (8-98) or an upgraded custom version (8-98A).

1931 Auburn Chassis Assembly. An Auburn innovation for 1931 was the addition of an "X" member to the frame. The "X" was a 6-inch deep channel of 5/32-inch steel with a spread of 75 inches, which provided exceptional structural rigidity. In what would seem to be an engineering step backwards, Auburn abandoned the use of hydraulic brakes in favor of mechanical brakes. Auburn's advertising agency coined the term "Steeldraulic" to describe the system.

1931 Auburn 8-98A Cabriolet. Prior to the late-year introduction of the Speedster and a seven-passenger sedan, the Auburn lineup for 1931 consisted of five body styles; a cabriolet, phaeton, brougham, coupe, and a five-passenger sedan. New for the year was an optional free wheeling unit. Attached to the rear of the transmission, it allowed the car to coast when the driver's foot was lifted from the accelerator pedal. This feature, of dubious value, caused extra wear to the already overtaxed mechanical brake system.

1931 Auburn 8-98 Engine. Built by Lycoming, Auburn's sole engine offering for 1931 was a straight eight-cylinder cast enbloc. It had a bore and stroke of 3 and 4 3/4 inches respectively. Displacement was 269 cubic inches. Horsepower was rated at 98 at 3,400 rpm. The crankshaft rotated on five main bearings, lubricated by a full pressure system. Aluminum alloy pistons were used. Other features included a Purolator oil filter, an updraft carburetor, and a 5.26 to 1 compression ratio.

1931 Auburn 8-98A Custom Sedan. The four-door sedan was by far Auburn's best selling model in 1931. A total of 15,131 examples were sold, comprising over 46 percent of Auburn production. The car shown here is a prototype photographed in 1930. It carries a straight front bumper, which differs from the production drop-center design.

1931 Auburn 8-98 Brougham. Auburn had not offered a two-door sedan type since 1926. The standard five-passenger brougham listed for $995 which placed it in the same price range as Oldsmobile, DeSoto, Hudson, and Nash. Stinson, a subsidiary of the Cord Corporation, manufactured the monoplane aircraft. The car pictured is a prototype.

1931 Auburn 8-98 Brougham Interior. The instruments were aircraft-type with a black background and white figures and indicator hands. Crowned glass lenses covered the dials. Controls on the panel included manifold heat, throttle, spark advance, ignition lock, lights, choke, and starter button. The hand brake operated on all four wheels.

1931 Auburn 8-98 Sedan. This lucky family accepts the keys to their brand new Auburn Sedan. They were winners in a raffle in which Auburns were given away to ticket holders who patronized participating Chicago merchants. On the storefront, Leg of Lamb and Fancy Chicken is advertised at 25 cents per pound.

1931 Auburn 8-98A Seven Passenger Sedan. Stinson Aircraft and Century Air Lines were an integral part of the industrial empire built by transportation magnate E. L. Cord. Auburn created the seven-passenger sedan to carry people between the airports and downtown areas. An attentive chauffeur awaits the arrival of his fares while parked near a Stinson Tri-motor passenger plane. The model shown is equipped with a specially constructed luggage rack on the roof.

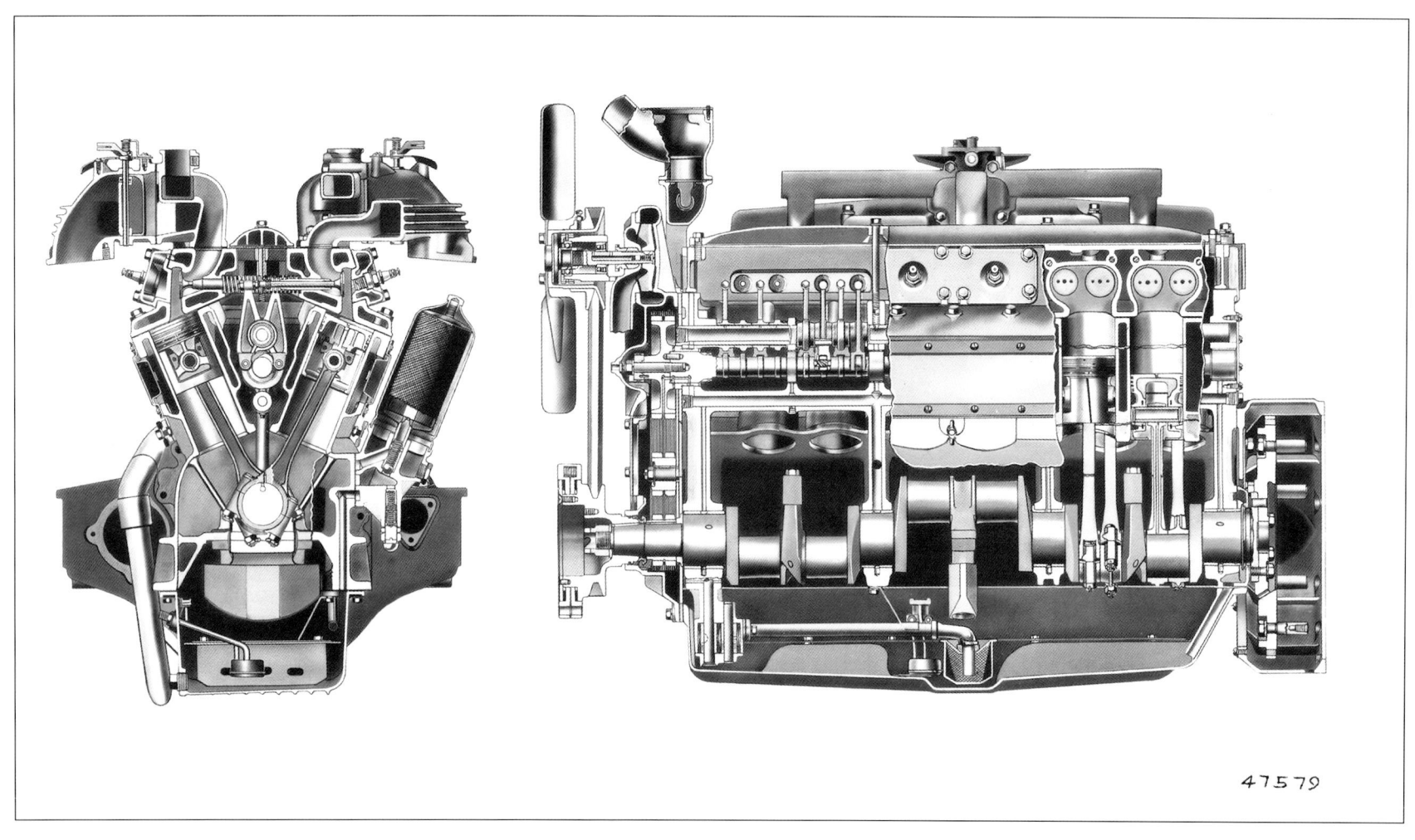

47579

1932 Auburn V-12 Engine. Lycoming introduced a V-12 engine in 1932, debuting in Auburn's new 12-160 Series. It had a 3 1/8-inch bore and 4 1/4-inch stroke yielding a lusty displacement of 391 cubic inches. Horsepower was rated at 160 at 3,400 rpm. Compression ratio was 5.75 to 1. To reduce overall width in order to fit Auburn's narrow engine compartment, the angle between cylinders was set at 45 degrees. A unique camshaft, rocker arm, and valve arrangement resulted in an unusual combustion chamber shape. Fins in the aluminum alloy oil pan provided additional oil cooling.

1932 Auburn V-12 Display Chassis. On display in the factory showroom is a Lycoming V-12 engine installed in an Auburn chassis. Visible in the photograph are twin ignition coils and the Bijur chassis lubricator located to the right of the exhaust pipe. The engine used two Stromberg carburetors although only one is visible from this angle. The Startix unit above the starter motor was a device that started the engine by merely turning on the ignition switch. This innovation restarted the engine automatically in the event of a stall.

1932 Auburn 12-160 Cabriolet. The Auburn Cabriolet had room for two up front with space for an additional passenger or two in the rumble seat. The V-12 models rode on a 133-inch wheelbase. The bumpers and knock-off hubcaps carried the twelve-cylinder logo to inform onlookers that this car was indeed special. Lycoming spent over $1 million in engineering and tooling to manufacture the new V-12 engine. Series 12-160 cars were equipped with hydraulic brakes while the eight-cylinder models continued with mechanical brakes.

DIAMONDS $1,000,000 LB

SHOES $5.00 LB

STEAK .35¢ LB

COFFEE .30¢ LB

AUBURN 8-100 SEDAN 25.3¢ LB

AUBURN SEDAN

3725 lbs. $945 F.O.B.

1932 Auburn 8-100 Sedan. Auburn sales sagged in 1932. This display in a Chicago dealer's showroom compares the price per pound of an Auburn Sedan to the costs of other consumer products. Drastic price reductions of $170 to $640 were made in June to help move unsold stock. The price of an 8-100 Sedan was cut to $775. A Stinson aircraft appears in the background.

1932 Auburn 8-100 Speedster. The air-conditioned Roosevelt Theater located in Chicago's loop was part of the Balaban and Katz Publix Theater chain. The theaters helped promote Auburn automobiles by sponsoring a giveaway raffle of 10 cars. Playing at the Roosevelt on this summer evening was Night Court, a sensationalized crime drama for "adults only." The price of admission was just 25 cents.

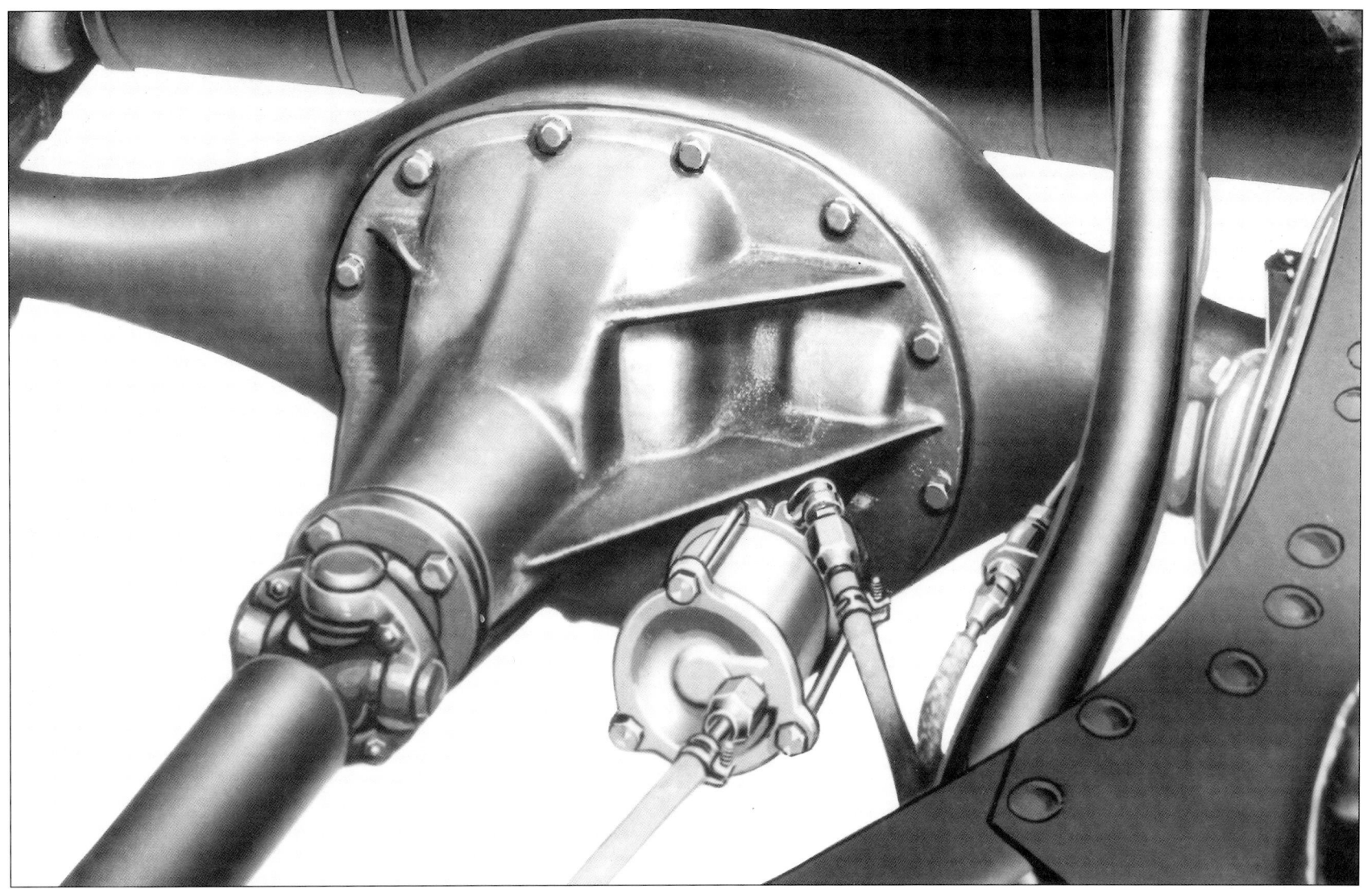

1932 Auburn Dual-Ratio Differential Unit. Manufactured by Columbia Axle Company, a Cord Corporation subsidiary, and introduced in 1932 was an optional two-speed rear axle termed "Dual-Ratio." Drivers could select a low or high rear end ratio in each of the three speeds offered by the transmission making Auburn the first six-speed automobile. A planetary gearset in the differential was controlled by a vacuum cylinder and valve. By simply turning a knob on the dashboard, the driver could select a low 4.5 to 1 ratio or a high 3 to 1 ratio. Dual-Ratio permitted both performance and economy.

1932 Auburn 8-100A Sedan. Auburn participated in many "economy runs" to tout the virtues of their two-speed rear axle. An eight-cylinder Auburn made a run from Los Angeles to San Francisco and return, 895 miles, with an average fuel consumption of 18.8 miles per gallon. A twelve-cylinder Auburn made the same run averaging 16.5 miles per gallon.

1932 Auburn 12-160A Brougham. The 4.5 to 1 rear end ratio engaged when the Dual-Ratio selector was in the Low position, made Auburn a worthy contestant in hill-climbing competition. Two twelve-cylinder Auburns set eight hill-climbing records in the mountains east of Los Angeles where grades as steep as 22 percent were encountered. Pictured from left to right are racecar drivers Eddie Miller, Earl Cooper, and William Claus with the trophies won by Claus.

1932 Auburn 8-100A Seven Passenger Sedan. This spacious Auburn Sedan provided comfortable transportation between nearby cities. The Greenville/Dayton (Ohio) Transportation Company ran a convenient passenger line between Fort Wayne, Indiana and Lima, Ohio (a distance of approximately 60 miles). This model, the largest Auburn offered in 1932, rode on a 137-inch wheelbase.

1932 Auburn 12-160A Speedster. The things dreams are made of. A comely model poses with an equally stunning Auburn V-12 Speedster. After the price cuts in June, this car could be purchased for $1,275. Only 35 V-12 Speedsters were built during the 1932 model year making this car one of the most rare of Auburns.

1932 Auburn 12-160A Speedster. The dramatic boattail shape of the curvaceous Speedster body is emphasized in this rear view photograph. The use of contrasting colors gives the car a more striking appearance. All twelve-cylinder Auburns came equipped with dual exhausts and two taillights. The top, when lowered, was completely hidden by a cover.

1932 Auburn 12-160A Speedster. Auburn once again proved its high-performance capabilities by shattering all stock car records for speed from 1 to 500 miles at Muroc Dry Lake in California. A Brougham and a Speedster, both using V-12 engines, established the new records for the American Automobile Association Class B (305 to 488 cubic inches). The Speedster, driven by veteran racer Eddie Miller, upped the flying mile benchmark to 100.77-mph, surpassing the old record by nearly 9 mph.

1932 Auburn 12-160A Phaeton. Three gentlemen admire the features of this handsome V-12 Phaeton with Grant Park and the timeless Chicago skyline in the background. This Auburn sports a rare factory accessory, a "Winged Mercury" hood mascot. The Phaeton and Cabriolet models had folding windshields. All eight and twelve-cylinder Auburns are recognized as "Full Classics" by the Classic Car Club of America.

1933 Auburn 8-101 Sedan Prototype. Placed on the turntable in the Auburn engineering department is an 8-101 standard five-passenger Sedan. Standard models used black painted lamps and wooden artillery-type wheels. The Sedan weighed in at 3,725 pounds and was priced at $845.

1933 Auburn 8-101A Sedans. In 1933 and 1934 Chicago played host to the Century of Progress Exposition where the Cord Corporation presented exhibits in the Travel and Transportation Building. Auburn sales continued to decline in 1933 and less than 5,000 units were manufactured. In spite of this, Auburn continued to pay dividends to stockholders. At the end of the fiscal year in November, a net loss of over $2.3 million was reported.

1933 Auburn Salon. Debuting in 1933 was the top-line Salon series with both eight and twelve-cylinder models. Apparent in this detail photograph is the new V-shaped grille and convex lenses on the headlamps unique to the Salon series. The redesigned radiator shell had no filler cap or name badge.

1933 Auburn 12-165 Salon Cabriolet. New for Auburn in 1933 for V-12 Salon models was a power braking system. A dashboard switch with settings for "Dry Weather," "Rain," "Snow," and "Ice" controlled the strength of the vacuum booster. Factory literature claimed, "So effective is the system that the brake pedal can be depressed with one finger."

1933 Auburn 12-165 Salon Sedan. A stylish front bumper and fender line distinguished the Salon models. Note the new two-piece V-shaped windshield that was a feature of the Sedan and Brougham Salon models. The Lycoming twelve-cylinder engine was unchanged for 1933.

1933 Auburn 12-165 Salon Brougham. The right-profile view illustrates the V-shape of the Salon windshield. The windshield could be opened for ventilation by operating a lever mounted on the top of the dash panel. The Brougham had a five-passenger capacity with its extra-wide rear seat. The front seats were individually adjustable. All Salon models were factory equipped with the Dual-Ratio axle.

1933 Auburn V-12 Salon Dashboard. The indirectly illuminated aircraft-style instruments included a combination ammeter-oil pressure gauge, heat indicator, speedometer, electric clock, and combination oil and fuel level gauge. Below the instruments are controls for manifold heat, ignition, choke, Dual-Ratio switch, throttle, spark advance, and lights. On the left is the brake power-booster adjustment lever.

1933 Auburn V-12 Salon Sedan Rear Interior. The Auburn sales brochure said of the rear compartment, "The upholstery is of high-grade broadcloth, pleasingly harmonized in color treatment with car exterior. Form-fitting luxura-type cushions. Garnish molding around windows, of two-tone Burl Walnut. Dome and reading lamps edged to match. Curtains are provided on the rear windows."

1933 Auburn 8-105 Salon Cabriolet. An owner-customized Auburn Cabriolet carrying a 1935 license plate is photographed at a Texaco service station in Chinatown in Chicago. Note the pagoda-style roof on the building. The Auburn has a modified grille and radiator shell in addition to several bolt-on items. Attached to the grille are images of dragons.

1934 Auburn 1250 Salon Cabriolet. Motion picture star James Cagney smiles for the camera in his new Auburn Cabriolet on the Warner Brothers lot. Left over 1933 V-12 models were rebadged as 1934s as Auburn sales continued to sag. The 1250 Salon Cabriolet with Dual-Ratio carried a factory list price of $1,695 in January of 1934.

1934 Auburn Styling Model. This model, made early in the development of the 1934 line, has few styling features that actually reached production. The unskirted fenders, oval shaped rear window, split rear bumper, and headlamp bracket were all rejected ideas. However, the belt molding, with minor modification, was one feature that did reach production.

1934 Auburn 850 Custom Sedan. Herb Snow (right), Auburn Vice-president of Engineering and a colleague take a break while on a long distance test run in a 1934 Sedan. The Auburn Automobile Company was too small to afford their own test track facility and thus did much of their testing on public roads. Unlike the car in the photo, standard production Auburns were equipped with wire wheels.

1934 Auburn 652 Standard Brougham. For 1934 Auburn introduced the 850 and 652 Series, both featuring welded all-steel body construction. The 652 Series was powered by a new six-cylinder engine of 210 cubic inches rated at 85 horsepower, Auburn's first six-cylinder since 1930. The five-passenger two-door Brougham was available in both Standard and Custom models. Canvas backdrops were used by factory photographers to aid advertising retouchers in airbrushing the background a solid white.

1934 Auburn 850 Custom Cabriolet. Auburn offered two eight-cylinder engines in the 850 Series. The Standard engine used a cast iron cylinder head with a 5.3 to 1 compression ratio and a single-barrel carburetor. It produced 100 horsepower. The Custom models were equipped with a 115 horsepower engine that used an aluminum cylinder head with a 6.2 to 1 compression ratio and a two-barrel carburetor. Both engines displaced 280 cubic inches. By now, opalescent and pearlescent paints were common.

1934 Auburn 850 Custom Phaeton. A chauffeur and his passenger pose for the camera in front of an Auburn dealership. The Phaeton was available only as a Custom model and the 850 listed for $1,225. The Phaeton and Cabriolet both featured fold-down windshields. Sales of the new 1934 Auburns showed a modest improvement over the previous year, however, the company continued to lose money.

1934 Corbitt Truck with Auburn Sheetmetal. Little is known about this handsome Corbitt truck fitted with 1934 Auburn body panels. Auburn automobile styling translated fairly well into truck duty. Corbitts were manufactured in Henderson, North Carolina from 1913 through 1958.

1934 Auburn Final Inspection. Factory workers at the Connersville, Indiana assembly plant install accessories and put the final polish on these 1934 Auburns. Auburn production capability was nearly tripled when the Connersville assembly plant opened in January of 1929. From 1934 on, all production was carried out in this more modern 81-acre facility.

1934 Auburn 850 Custom Sedan. A Chicago policeman escorts school children across the street in this staged safe driving demonstration. Auburn's hydraulic brakes with 12-inch diameter drums provided adequate stopping power. The shape of the lower grille opening suggests that Auburn engineers may have considered front-wheel drive. The 850 Custom Sedan listed for $1,125.

1935 Auburn 653 Sedan. Body designer Gordon Buehrig was brought from Duesenberg in February of 1934 to restyle the 1935 Auburn. His challenge was to give the car a more powerful, racy appearance which he accomplished by redesigning the grille, hood louvers, and front fenders. The "Winged Goddess" image was used as a hood mascot. A Lycoming six-cylinder engine powered the 653 models. Wheelbase was set at 120 inches.

1935 Auburn 851 Salon Phaeton. Phaetons offered the best of both worlds. During foul weather the snug fitting top and roll-up windows provided occupants protection from the elements. In fair weather with the top and door windows lowered and the windshield folded flat, open-air touring was a pleasant experience. The 851 models used a Lycoming eight-cylinder engine and rode on a 127-inch wheelbase.

1935 Auburn 851 Brougham. "Room to spare for five full-grown people in this Brougham, yet it is very compact. The two doors are exceptionally wide, making getting in and out of the rear seat perfectly effortless. Upholstery is of high quality broadcloth," so said the 1935 Auburn sales brochure. The bumpers on this example are fitted with factory approved guards.

1935 Auburn 851 Coupe. The buyer of a Coupe had a choice of a roomy baggage compartment or, at additional cost, a rumble seat for extra passengers. The Coupe was constructed by adding a fixed padded solid top of wood framework to the basic Cabriolet structure. The list price of this model was $1,045.

1935 Auburn 851 Landau Sedan. A one-off effort, the Landau Sedan featured luxury interior appointments including a curved divider window between the rear passengers and the driver. The landau irons were non-functional and the roof was fixed.

1935 Auburn 851 Dashboard. On the left side of the panel the oil pressure and fuel gauges, water temperature indicator, and ammeter were grouped in a round dial. A 5,000-rpm tachometer occupied the center of the instrument panel on eight-cylinder cars equipped with Dual-Ratio. A 100-mph speedometer rounded out the instruments. In the center of the dash a provision was made for a built-in radio. An electric clock, standard equipment on the 851, was located in the center of the glove-compartment door on the right. The panel itself had a mahogany and chrome finish. The Dual-Ratio control was located on the steering wheel hub.

1935 Auburn 851 Sedan with Racer. The Official Car of the Midwest Auto Racing Association was this handsome Auburn Sedan. Photographed while parked outside the Auburn assembly plant in Connersville, Indiana, the rig is ready for racing with its "Auburn Special" in tow.

1935 Auburn 851 Speedster and Phaeton. Arguably the most dazzling Auburn ever built was the 851 Speedster (foreground). Gordon Buehrig's inspired design has endured for decades as a benchmark of beauty and grace. Hollywood film stars Mary Astor and George Murphy pose with a Phaeton and a supercharged Speedster. The cover of the Auburn sales booklet aptly stated, "Exclusive - Distinctive - Individual" in describing the 1935 line.

1935 Diesel Powered Auburn 655 Phaeton. The developer of the high-speed diesel engine, Clessie Cummins, promoted this car as a viable alternative to gasoline powered automobiles. Officially known as the 655 model, seven examples were reportedly built to demonstrate economy of operation. Mr. Cummins drove the car from New York to Los Angeles, averaging 34.6 miles per gallon of fuel oil. Total cost of fuel for the trip was a mere $7.63.

1935 Diesel Powered Auburn 655 Phaeton. Clessie Cummins points out the features of his diesel powered Auburn to a multitude of curious New Yorkers. The six-cylinder engine had a block and cylinder head made of aluminum alloy. The entire engine weighed only 80 pounds more than the Auburn eight that it replaced. It developed 85 horsepower at 2,200 rpm. High manufacturing costs doomed the diesel as a substitute for the gasoline engine in 1930s America.

1935 Auburn 851 Super-Charged Phaeton. E. L. Cord donated this car to the Indiana State Police to be used for safety education and enforcement. Paul Beverforden, the fortunate officer assigned to this vehicle, could outrun just about any speeder. The patrol car was later fitted with a red light, siren, and a fender-mounted loudspeaker.

1935 Auburn Super-Charged Eight-Cylinder Engine. To increase performance, Lycoming Manufacturing fitted their eight-cylinder Auburn engine with a Schwitzer-Cummins centrifugal supercharger that ran at six times crankshaft speed. The 6.5 to 1 compression ratio combined with the boost from the supercharger was sufficient to increase horsepower from 115 to a potent 150. A bore and stroke of 3 1/16 x 4 3/4 inches yielded a displacement of 280 cubic inches.

1935 Auburn 851 Speedster. In July of 1935 the Auburn racing team arrived at the Bonneville Salt Flats in an attempt to set new speed records. Crew members install a fresh set of Champion spark plugs in a stock supercharged Auburn Speedster under the vigilant eyes of an AAA official. Auburn smashed 70 official American stock car speed records, including the flying mile, at over 104 mph and 1,000 miles at nearly 103 mph.

1935 Auburn 851 Speedster. Renowned speed endurance driver Ab Jenkins was at the wheel during the speed trials at Bonneville. His instructions from H. T. Ames, executive vice-president of Auburn, were to "nail the accelerator to the floor and drive wide open - no mercy - no nursing - no let-up." Auburn's performance at the salt flats grabbed newspaper headlines throughout the country, but unfortunately, the publicity did not translate into increased sales. In August of 1935 Ab Jenkins would return to Bonneville to set even more records in a Duesenberg.

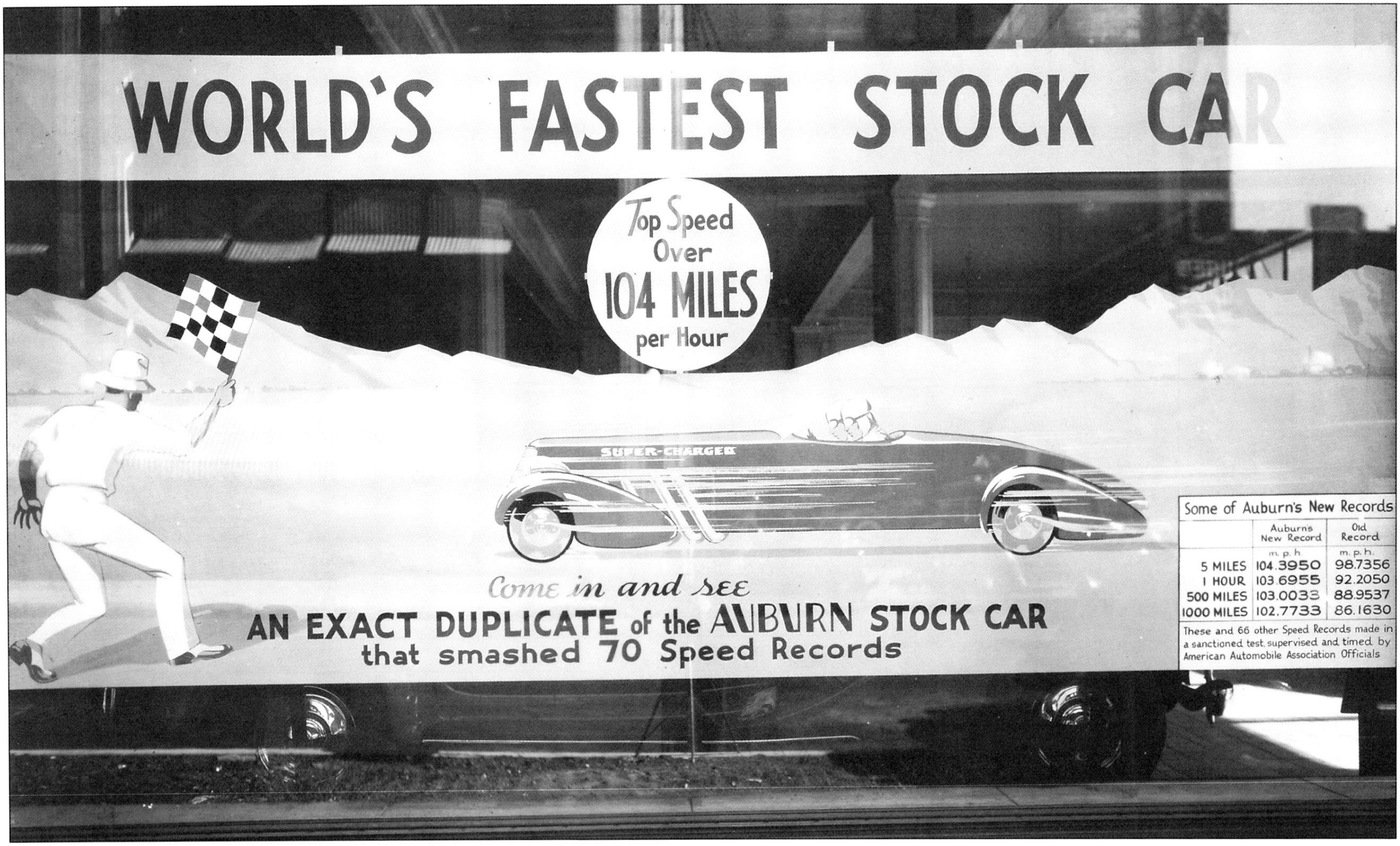
WORLD'S FASTEST STOCK CAR
Top Speed Over 104 MILES per Hour
SUPER-CHARGER
Come in and see
AN EXACT DUPLICATE of the AUBURN STOCK CAR
that smashed 70 Speed Records
Some of Auburn's New Records
Auburns New Record
Old Record
m. p. h.
m. p. h.
5 MILES 104.3950 98.7356
1 HOUR 103.6955 92.2050
500 MILES 103.0033 88.9537
1000 MILES 102.7733 86.1630
These and 66 other Speed Records made in a sanctioned test, supervised and timed by American Automobile Association Officials

1935 Auburn Dealer Showroom Window. Auburn attempted to make the most of its high-performance image as displayed in this dealer's window. Each Speedster left the factory with a dashboard plaque signed by Ab Jenkins certifying that the car had been driven at over 100 mph. As Auburn had no test track facility, public roads were used for such runs. Norbert Johnson, employee in the Experimental Department, later said, "The State Patrol raised the devil a couple of times. They said we (test drivers) shouldn't use the road for a race track. But, they never bothered chasing us because our cars were so much faster than what they had."

1935 Auburn 851 Speedster. Ab Jenkins is at the wheel of his Speedster mugging for the camera with singer Gene Austin and comedians Olsen and Johnson of "Goin' Places," a touring musical production. The troop of 50 stage, screen, and radio stars used six new Auburns and the bus in the background in their caravan. Note the fender skirt covering the rear wheel opening of Jenkins' car.

1935 Auburns - "White Caravan." A Sedan, Speedster, Phaeton, and Cabriolet are pictured in front of the Auburn Administration Building as they prepare to leave on a promotional excursion. During the summer and fall of 1935 teams of four lustrous, white Auburns piloted by drivers dressed in matching white outfits toured 150 dealers throughout the country. Known as the "White Caravan," they paraded through the business sections of cities and appeared at civic clubs and schools. Note the loudspeaker on the front fender of the lead car. The art deco style Administration Building is currently the home of the Auburn Cord Duesenberg Museum and is listed in the National Register of Historic Places.

1935 Auburn 653 Sedan. Members of the "White Caravan" stop by Landy's Clothing Store in Auburn, Indiana after being outfitted in swank-looking white suits.

1935 Auburn 851 Super-Charged Sedan. A dapper gentleman poses with a new Sedan in front of the Auburn Automobile Company Administration Building. The "Super-Charger," as Auburn called it, was an option on all eight-cylinder equipped cars (standard on the Speedster). Super-Charged Auburns were easily identified by four flexible stainless-steel exhaust pipes exiting the engine compartment through the left hood side and running through the fender.

1935 Auburn 851 Hearse. Late in 1935 Auburn expanded its lineup to include a hearse, ambulance, and a nine-passenger limousine. The wheelbase of these luxuriously appointed models was a lengthy 163 inches. The bodies were built in-house for these limited production vehicles. Two nine-passenger limousines were equipped with Cummins Diesel engines for potential use as airport and hotel shuttles.

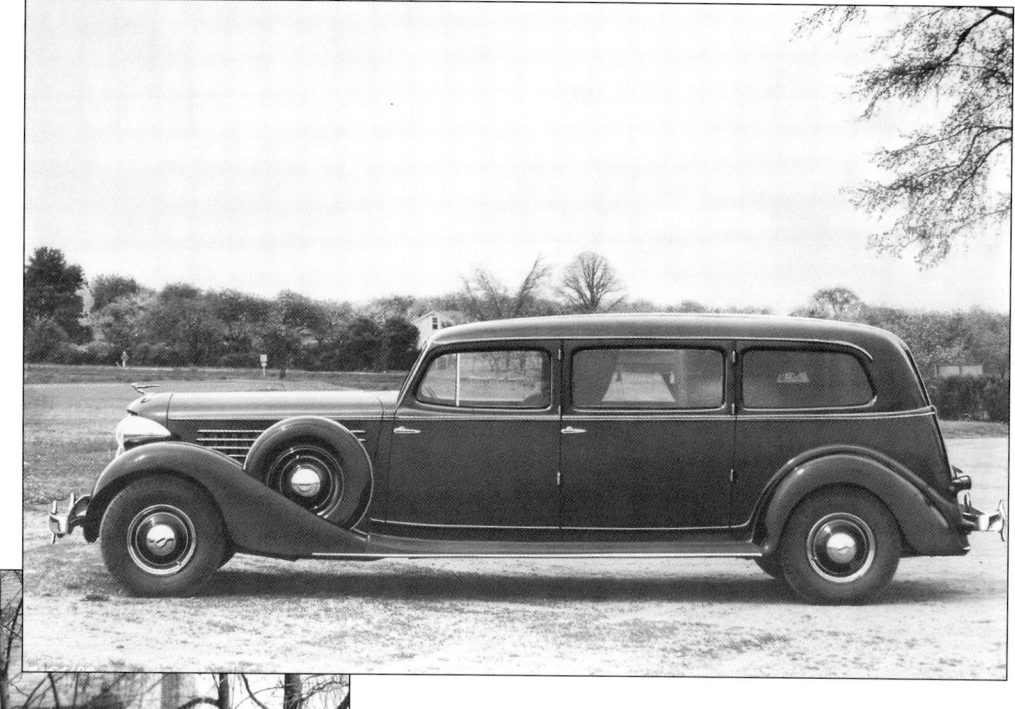

1935 Auburn 851 Sedan and Trailer. The Sedan's meager luggage space of 13.5 cubic feet caused Auburn to consider production of this stylish, single-wheel trailer. It was securely attached to the rear bumper of the car with two clamps instead of using a trailer hitch. The caster-style wheel was on a pivot to allow for turns and ease of backing up. The Sedan is pictured with accessory chrome wheel covers.

1935 Auburn 851 Special Super-Charged Coupe. This factory-built one-off Auburn Coupe has been tastefully customized with the addition of Speedster fenders, running boards, and a steel top. The use of unframed door glass gave the Auburn Coupe the sleek, airy appearance of a hardtop convertible.

1936 Auburns – Export. Auburn's Export Division, headed by the globetrotting Robert Wiley, had dealers located in 99 countries throughout the world. Vehicles, partially disassembled, were shipped abroad in wooden crates. These Auburns were bound for the Autohandel Marcusse dealership located in Medan, Sumatra in the Dutch West Indies.

1936 Auburn 654 Cabriolet. Many Auburns sold as 1936 models were unsold, rebadged 1935s. The six-cylinder engine used in the 654 had the same bore and stroke as the eight allowing for greater interchangeability of parts. The 654 Cabriolet pictured is parked in the driveway of "Cordhaven," E. L. Cord's 62-room mansion in Beverly Hills, California.

Fuller Motors Auburn Dealership circa 1936. Sales plunged to 1,250 units in 1936, the final year for Auburn. Late in the year, dealers heavily discounted leftover Auburns. Fuller Motors, located on Wilshire Boulevard in Los Angeles, was offering price reductions of up to $500. As late as 1937 there remained nearly 150 unsold Auburns on factory lots. The Auburn Automobile Company was now concentrating its efforts on the newly introduced 1936 Cord. The radio transmitting towers atop the building belonged to E. L. Cord's broadcasting station.

1908 Auburn Model H Touring. It is an uphill battle for an independent automobile manufacturer to survive in the United States. Auburn gave all it could to endure. The company always provided a product of exceptional style, innovation, performance, and value. However, the cold-hearted economic realities of the industry dictated that the extra financial burdens placed on the small manufacturer were sufficiently heavy to cause eventual extinction.

MORE TITLES FROM ICONOGRAFIX

All Iconografix books are available from direct mail specialty book dealers and bookstores worldwide, or can be ordered from the publisher. For book trade and distribution information or to add your name to our mailing list and receive a **FREE CATALOG** contact:

Iconografix, PO Box 446, Dept BK, Hudson, Wisconsin, 54016   Telephone: (715) 381-9755, (800) 289-3504 (USA), Fax: (715) 381-9756

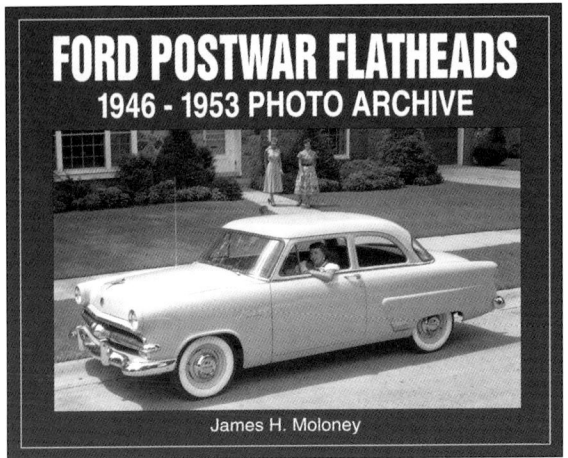
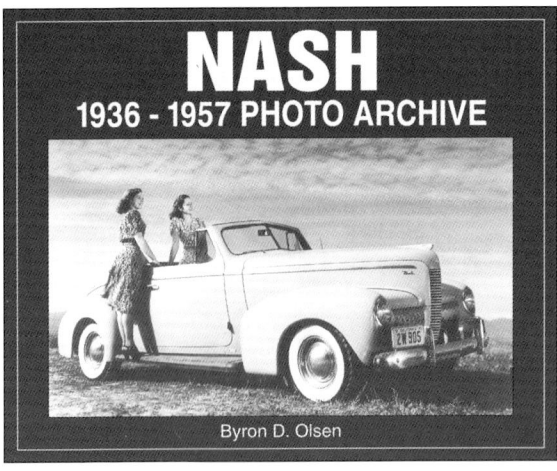
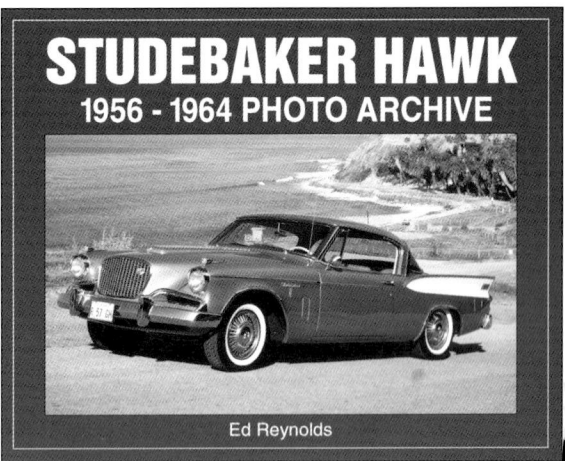